Animal emotions

To Tennessee, Aliosha and Ilona.

Konecky & Konecky
72 Ayers Point Rd.
Old Saybrook, CT 06475

ISBN 1-56852-581-8

Translated from the French by
Rosetta Translations
121 Godolphin Road
London W12 8JN, England

Printed in China

Karine Lou Matignon

Animal emotions

KONECKY&KONECKY

"Animals live in
and of sensory pe
are capable of aff
suffering but, for al
the same as human
paradox is that the
origins of our ow
animality that stil

world of emotions

ception. They

tion and of

hat, they are not

beings. The

each us the

behavior, the

ies within us."

Boris Cyrulnik

Emotions are shared

As a child I used take refuge with animals to escape from the world of human beings, their contradictions and the arrogance with which they believed that they were exceptional. I would have loved to be an animal and remain wild. Nature sharpened my senses and my rebellious temperament before civilization caught up with me. At first this was through fairy tales and myths, in which the animal symbolically telling the story of the life of humans and of gods began to tame me. Later it was through science, where the scientists deciphering life and revealing our animal origins deeply disturbed me. I have been fortunate enough to meet many passionate men and women: artists, thinkers and scientists who have answered my questions. They have allowed me to share their experiences and their fascinating researches, and in this way they have reconciled me with this world.

For a long time, it was easy for us to believe that we were the elect, reducing animals to mere mechanical objects-a practical way to protect ourselves against any criticism! What would happen to our justification for exploiting animals if it turned out that they were not machines? In 1872, the naturalist Charles Darwin paved the way when he published his book The Expression of the Emotions in Man and Animals. He believed that animals are capable of feeling and expressing suffering, pleasure, happiness, misery, fear, loyalty and jealousy, and that the absence of language was no proof of any discontinuity between animals and men. The brain that controls anger, fear and desire (the limbic system) is relatively simple in fish and reptiles, but in birds and mammals it is very close to ours. This suggests that many species can feel the same emotions as we do ourselves.

Employing scientific data, moving stories, telling anecdotes and above all remarkable photographs, this book should be read as a plea on behalf of animals, revealing them as they really are, neither objects nor really human, but sensitive beings with awareness, intelligence, language and emotions of their own. Respecting and understanding animals without going too far also means respecting humans, understanding their place in nature and avoiding the creation of a hierarchy among living beings. From the moment we begin to understand how another feels, barriers disappear and differences are no longer alarming. Animals can be seen on their own terms. It has long been thought that to be interested in animals is to turn one's back on people, but I believe that, on the contrary, knowing and understanding animals can help us develop more fully as human beings. It also helps us to become aware of the impact of our behavior on our surroundings and so to protect our natural environment that is so fragile and yet so important to our survival.   K. L. M.

Animal passions

Human beings have no monopoly on amorous passion. Animals do more than merely carry out nuptial rituals driven by simple physical urges at certain times of the year.

Beyond the value of love in the process of selection, the strategies of seduction together with the alchemy of desire and pleasure also expand the amorous animal's emotional repertoire.

Hyper-
sensuality

Previous double spread: The female dolphin
presents her belly, entering into the ritual
dance that precedes mating.

There are some species that can fertilize
themselves (this is the case with wasps and
aphids) and others, such as certain fish of
the Great Barrier Reef in Australia, that have
the unusual ability to change their sex. If
the male disappears, the dominant female
metamorphoses into a male! Sexual
reproduction is not the only process
invented by evolution, but it is the
predominant means of ensuring the
multiplication of the species while
encouraging new associations of genes. But
for mating to take place, there must first be
a couple, and to achieve this, nature has
proved itself to be extraordinarily
imaginative: songs, scents, offerings,
synchronized dances and sensual caresses
have been invented. Giraffes, whales, moles
and lions rub tenderly against each other
before mating, while elephants stroke each
other affectionately with their trunk for
hours on end. Unlike the majority of species,
dolphins deliberately choose their partner
and make love throughout the year. Outside
the mating season, they continue to live a
life of sexual gratification. For instance, it is
not unusual to see couples stroking each
other while making little noises, whispering
and whistling. They nibble each other, kiss,
bill and coo, and explore each other's bodies
and genital organs with their snouts.

The hormones
of love

Opposite: As soon as he has found a female,
the male bear courts her by nibbling at her neck.

Animals express their emotions through
posture and vocalization, as well as through
the odors and scents that play an important
part in the formation of couples. With the
arrival of spring, many male birds (ducks,
blackbirds, pigeons, etc.) are completely
bewitched by the signals given out by
females, altogether forgetting to be vigilant
and in many cases ending up in the mouth
of a predator or under the wheels of car.
What is the source of this attraction that is
so powerful that it annihilates everything
else? What substance is it that causes this
amorous passion? It is part of a family of
more or less volatile organic molecules
known as pheromones (from the Greek
pherein hormân, meaning "to carry
excitement"). These are hormonal
substances produced by the glands and
secreted outside the body. They are received
by the olfactory senses and modify the
behavior of those who breathe them. They
are used for communication between
members of each species, enabling them to
identify each other and to mark their
territory. They also play a role in all kinds of
behavior including sexuality.

    The male bear attracted by the
scent of a female immediately goes in
search of her.
In societies where hierarchy establishes a
dominant couple, the estrogen hormones
present in the dominant female's urine have
an inhibiting effect on the ovulation of the
other females (this is the case among wolves
and mice). The legacy of a biological animal
past, love at first sight in human beings can
also be triggered by one of the most
primitive senses: the sense of smell.

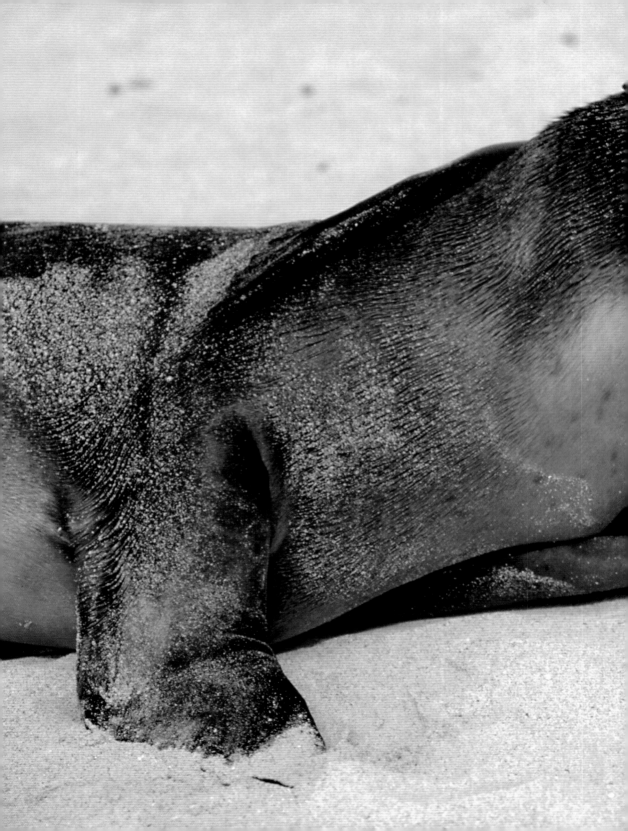

The
amorous brain

Previous double spread: A male sea lion can set up a harem of seven to ten females and remain on and without eating for two months while he waits for all his females to be fertile.

Although emotions are not simply the result of a mixture of hormones, the part hormones play in an animal's emotional state, including desire, remains important. The production of hormones is based in two glands situated at the base of the brain: the hypothalamus and the pituitary gland. Also known as the centre of emotions, the hypothalamus controls sexual behavior. Under the influence of the environment (a change in temperature or the amount of light, a specific temperature, conditions of social life, etc.), it produces a substance that is transmitted to the pituitary gland. This is an almond-sized gland that triggers milk production and uterine contractions, but also the production of the hormones responsible for desire and for social and sexual satisfaction and pleasure. A castrated female monkey no longer producing attractive pheromones will no longer be of any interest to her partner, unlike the rutting male elephant whose temples ooze a thick liquid that stimulates the females. Whatever the species and the type of mating, it has been shown that males and females have certain preferences in their choice of partner. But guaranteeing the survival of the species is not the only objective: leopard seals, mouflon sheep, and bison are among the animals that seduce each other and also mate for pleasure with other members of their species of the same sex. For a long time, homosexual behavior among animals was considered abnormal by naturalists.

Dancing
in unison

Opposite: In performing their nuptial dance, cranes can jump more than six-and-a-half feet off the ground.

When trying to attract each other, many birds engage in highly ritualized dances. Sometimes, for instance with cranes this dance involves a remarkable degree of bodily synchronization. The male swan mirrors his partners dance, reproducing the smallest movement made by his companion, thus seducing and bewitching her. Cats walk round and round, deer leap up and down, chimpanzees swing without taking their eyes off each other, and howler monkeys favor jerky movements of the tongue. To express their desire, brown bears, wolves and foxes carry out a sort of rhythmic dance lasting several days. This ritual theatrical performance has several functions. It demonstrates the male's ability to invest in a relationship, it displays the color markings of the body (a measure of good health), it enables the animals to find their ideal partners by osmosis with their internal rhythms, and it increases their ability to communicate and to be aware of each other by stimulating the senses.

Unconsciously miming the movements of a partner and responding to them is a universal phenomenon that is also found in human beings. In the ceremonial ritual of the couple forming, the animals' expressions and body language also provides a channel for aggression, inseparable from sexuality. This is why, when courting, black-headed gulls, albatross, ganders, horses and cats turn their heads or eyes away when approaching their partners. In this way each is trying to hide the signals that at other times would lead to fear and aggression.

Wedding presents

Previous double spread: With a wingspan of ten to thirteen meters, the albatross spends long periods over the sea (sometimes as long as eight years) and only returns to land to mate.

To seduce a lady friend before the act of love by giving her little presents is a sexual stratagem that is found as much among birds and mammals as it is among human beings. According to experts, it is a custom that probably predates the dinosaurs because of its association with reproduction. Gift-giving males also demonstrate their hunting skills and their ability to feed their young. Thus, in the course of their amorous dance, great crested grebes offer each other algae while dancing an elegant little pas de deux, also known as the "penguin step," while the male falcon brings the female food in full flight. In the case of terns the male offers the coveted female a small fish, while the California roadrunner will choose a lizard. The wandering albatross brings no gift apart from its amazing patience! It is only after three or four years of complex rituals involving tickles, calls, clattering of the beak, elaborate dances and courtly bows that couples finally pair together for life.

Sexual animals

Following spread: When the cheetah has identified the presence of a female in the distance, he communicates with her by making noises.

Each species has its own sexuality. Gibbons are monogamous, orangutans are solitary creatures, while male and female macaques mate with several partners and maintain a special relationship with them even outside the mating season. Male hamadryad baboons keep their females in a virtual harem that requires their constant vigilance. Young male rivals prowl around the area where the clan is based, using humor and play, an infallible tactic for seducing the females. The bonobos who live in the swampy jungle of the Congo have sex every day. They exchange long kisses with their tongues, they embrace, they make love in all positions, including frontal, and they practice fellatio and other genital caresses. They are in search of pleasure, but not only that. Bonobos practice use these variations of free love to calm tensions that might otherwise divide the group, to strengthen the ties between them and to ensure better sharing during meals. By bartering an embrace for a banana or a stem of sugar cane, they avoid aggressive behavior. Although up to eight males are capable of queuing patiently as they wait their turn to copulate with the same female, there are others who manage to secure an exclusive relationship.

In the laboratory, in 85% of cases it is the female chimpanzees who take the initiative to mate, but the males can also be quite persuasive. Sitting on the ground with legs apart, a male will not hesitate to display his erect penis while devouring the chosen female with his eyes in an attempt to persuade her to submit. Cheetahs are more demure in their amorous frolicking. During their restless wandering the nomadic females meet male cheetahs, sometimes brothers, who have joined forces to control a territory. After copulating the lovers go their own way.

Serenades
for nuptials

Previous double spread: The great crested grebe has developed an amorous language consisting of purring, cackling, and trumpet noises.

Research has revealed a certain similarity between the acoustic signals made by animals during the seduction phase and human conversation: couples chatter a lot to seduce each other! As soon as the days become longer in our latitudes, that is at the end of January, it is very common to hear birds break into little songs.

The longer hours of daylight trigger a signal in the brain marking the beginning of the mating season and intensify their sexuality by stimulating the secretion of hormones acting on the sexual glands. If these birds were blindfolded, thus depriving them of light, they would remain silent and indifferent to love! So far as seduction is concerned, singing is one of the strategies followed by both animals and humans.

Otto Jespersen, a philologist who studied ancient civilizations, put forward the hypothesis that language was born from the first cries of love. As he put it, "Language was born in parallel with the seductive behavior of humans and I imagine that the first verbal expressions were halfway between the lyric solos of tomcats on rooftops and the melodious love songs of the nightingale." In any case, whether among the cetaceans, gibbons or birds, it is the rhythm and power that seduces rather than the melody. In the case of canaries, the one whose song is the most strident, is telling the female very clearly that she is in the presence of a male in perfect physical condition.

A seductive gaze

Following spread: During the mating season pumas live side by side for ten days, sleeping and hunting together.

According to anthropologist Barbara Smuts, during the mating season baboons stare at each other intensely, and although this animal branch separated from the trunk of human evolution over 19 million years ago, baboons and humans still have a lot in common in their ways of courting. Indeed, the ethnologist Helen Fisher, specialist in matters of love and courtship, describes how a female baboon named Thalia began to look a young male, Alex, up and down until she caught his eye. "Immediately he flattened his ears against his head, lowered his eyelids, and began to lick his chops, a mark of affection among baboons. Thalia was transfixed. She continued to stare at him for quite a long time. It was only after a long exchange of looks that Alex approached her and she started grooming him: this was the beginning of a friendship and sexual relationship that was still going strong six years later." Even a predator such as a tiger or puma will gaze at his partner with the tender look of a seducer.

Faithfulness and promiscuity

Opposite: Giraffes live in herds that vary in their composition. Only the dominant males mate.

Having enjoyed their nuptials, foxes continue to enjoy a friendly relationship to such an extent that the male will sometimes bring back prey to his sick partner. After mating, the field vole remains near his female and continues to stroke her, smoothing her fur as he watches jealously over her. Italian scientists have shown that human behavior in love is based largely on the same neuro-biological foundations as that of the vole! Thus it has been shown that the production of two hormones in the brain, oxytocin and vasopressin, encourage family-orientated behavior, care for the young, paternal feelings and monogamous behavior in the male. Sexuality therefore also contributes to create social links that will encourage a helpful attitude and faithfulness on the part of both males and females, thus enabling them to bring up their young together. But not every creature manages to remain faithful. While 90% of birds form couples, only 3 to 5% of mammals (4,000 species including humans) favor a long-term partner. This is the case with jackdaws, swans, robins, some muskrats, Asian clawless otters, beavers, gibbons, foxes, coyotes and wolves. The crow, like the graylag goose, becomes distraught if its partner dies. Nevertheless adultery is not unknown. With squirrels, for instance, the male is so jealous that he keeps watch over the nest and keeps a close watch over his flighty partner, while the rat tends to lose interest in his partner as time goes by.

The ultra male

Following spread: Kangaroos organize themselves in groups of two to ten animals. Only the "braggarts" rise to the rank of dominant male.

Competition among male birds inspires elaborate song and plumage. As with human females-who according to serious research have a marked preference for rich, ambitious men-females in the animal kingdom also want to be assured that the potential sire will be equal to his responsibilities. They look for strong mates with symmetrical features, an indication of good health. Work on the impact of pollution on the development of animals has confirmed these findings. A deer with asymmetrical antlers, a cockerel with a comb that is too short or not bright red or imperfectly shaped, or a swallow with an excessively short, irregular tail will have little chance of finding a partner to mate with. It is well known why the peacock spreads his tail: the most flamboyant suitor will be chosen. This instinct is strongest among kangaroos where the dominant males are prepared to fight each other for the female. Birds of paradise dance to show off the brilliance of their plumage and American grouse color the pockets of naked skin hidden under their feathers red or yellow. For years experts were unable to persuade flamingoes to reproduce in zoos until they discovered what really excited the females: the bright orange flight feathers in the male's wings! In the wild, this color is the result of the flamingoes eating large quantities of brine shrimp larvae (Artemia), the pigments of which are absorbed into the blood, thus distributing them throughout the bird's body. The food given in zoos did not have these pigments so the captive flamingoes lacked this exciting color. The mystery was solved.

A mother's love

Most scientists do not believe in
maternal love among animals.
They see only behavior patterns
that ensure the survival of the species,
the result of particular reflexes and
conditioning. Yet mothers look after
their young with care and tenderness,
and for their part the young seem
really attached to their mothers.
Are these acts completely devoid of
feeling and, if these emotions exist,
what are they based on?

A boundless attachment

Opposite: Canadian scientists have exposed the trauma and serious distress suffered by cows when they are separated from their calves after birth.

The harmony that exists within a herd of cows is based on rules controlling social life. These rules are based on the strong links that exist between a mother and her young. Demonstrating this is the story of a two-year old English heifer named Blackie, who fled from the farm to which she had just been sold in order to find her calf about six miles away, in an area that was completely unknown to her.

Reported in the daily World Farming Newsletter, the story began when the cow and her calf were sold at the market in Hathersleigh, a delightful little town in Devon, in the south of England. The cow was bought by one farmer and her calf was acquired by another breeder. Well fed and accommodated, the cow nevertheless decided to escape by jumping over the hedge. She was discovered and identified the following morning, quietly feeding her calf several miles away.

Maternal memory

Opposite: The baby orangutan will remain with his mother for seven years. If separated from his mother, the traumatized baby could die of his despair.

Following spread: Lionesses hide their young to protect them from the attacks of males who are not part of the clan.

During a trip to Botswana in Africa, the French elephant expert Pierre Pfeffer, witnessed the reunion between a matriarch and her son, who had not seen each other for years. As they were members of different herds, the meeting happened near a watering hole. When the young male elephant came closer, the mother suddenly left her herd and ran towards him, trumpeting with joy.

From the moment of birth, the newborn and the mother become attached to each other. In some species the mother identifies her offspring by licking it; the amniotic liquid contains distinctive chemical molecules. Consequently the cow can recognize her daughter until she in turn gives birth, in spite of any separation that may occur in the hours and days following the birth. If the memory of smell is so persistent, the visual memory is just as much so. Researchers at the Babraham Institute near Cambridge have shown that sheep are endowed with a complex intelligence, emotional reactions, and the ability to recognize other familiar fellow-sheep, remembering their faces for up to two years. They have identified in sheep the same specialist visual cells used for the recognition of faces as those in human beings and large primates who are able to recognize members of their families in photographs. On the other hand, lionesses, like cats, recognize, feed, and protect the young of all the females in a group indiscriminately.

Privileged connections

Opposite: During the first weeks of life, the female panda rocks her baby non-stop.

As soon as mating has taken place, then again at the birth, the neurons (nerve cells in the brain) secrete the hormones prolactin and oxytocin (the hormone of love), which trigger the production of milk (or the laying of eggs in birds), as well as encouraging and ensuring that care is taken of the young. Grooming and feeding her young from the moment of birth ensures that the mother becomes attached to her offspring while awaking and stimulating its development through her scent and thus establishing a privileged relationship between them. From the moment of her baby's birth, the female panda protects, calms and fusses over her young. As in all species of mammals, the earlier the attachment, the stronger it is. Researchers believe that, like sexuality, the development of maternal tenderness and attachment is triggered by environmental factors, combined with the production of hormones that influence behavior. The best known example is that of the young seagull, excited at the sight of the red dot on the parent's beak. By pecking at it, the young bird also stimulates the feeding reflex.

In the same way, wolves regurgitate their meal to give to their young as soon as they show they are hungry. None of this suggests that this parental behavior is prompted by anything other than a desire to promote the well-being of the young. It may happen that a mother becomes stressed during her own development or emotionally disturbed during the gestation period in a way that can prevent the bonding between mother and young. This behavior may occur in the wild but it happens most often in captivity where conditions do not make the animal feel secure. The mother then will abandon her young.

Clamoring
for love

Following spread: Baby penguins recognize their parents' song after only a few seconds and this in spite of all the noise around them.

Sound plays an important part in the parents' recognition of their young. In penguin colonies that sometimes consist of over one million individuals, birds recognize their partner or young on the basis of their acoustic identity. If the birds' ears are plugged with wax, members of the same family can pass each other without recognizing each other. It has been shown that turkeys experimentally made deaf no longer recognize their young and may even kill them.

Similarly, a chicken separated visually from its chick will remain interested in her young because of its chirping but if the chicken is prevented from hearing the chick, she will lose interest in it even if she can see it.

Love
is blind

Opposite: If separation occurs at particular moments during the development of a young albatross, the bond between it and its parents may be destroyed.

Living beings can only develop properly in a familiar environment. Between the 13th and 17th hour in the life of a chick, everything and everyone coming in its field of vision will arouse its interest. Whether it is a bicycle, another animal or a human being, it will associate this image with its familiar world and become attached to it. This is how impressions become imprinted in the chick's awareness, the moment when the newborn immerses itself in its mother's presence.

It was the Austrian Konrad Lorenz, one of the pioneers of ethology, who in the 1950s contributed to the discovery of this period of imprinting thanks to a little graylag goose, called Martina. Barely out of the egg, the gosling adopted the ethologist, preferring him to any other adoptive mother. "I had not understood the seriousness of the responsibility I had taken on when I looked in those little black eyes," he said, "and greeted it ceremoniously with a few words [...]"

Culture
and tradition

Previous spread: Every group of chimpanzees has its own technique for cracking nuts or hunting ants that is different from that of others.

The longer the development and the later the maturity of an animal, as is the case with elephants, large primates or killer whales, the longer a mother will care for her young. With killer whales, society is based on matriarchy. A dominant female is surrounded by her family, which enables the group to develop collective knowledge such as learning to hunt. Gently, the mother pushes her child towards the beach to teach him to catch an eared seal, in the same way that a domestic cat or a big cat will bring a mouse or young antelope to her young to teach them how to kill. Through imitation, one of the keys to survival, and the observation of their fellow creatures to which they are attached by social ties or bonds of affection, social birds and mammals share traditions and pass them down. A young panda will learn the secrets of successful mating by observing adults reproducing themselves.

Brought up in a calm atmosphere, emotionally well balanced, a young chimpanzee will effortlessly find the right place in the social group and will very easily memorize the customs of the clan that have been passed on-such as how to crack nuts or catch ants using a particular technique.

Adoptive parents

Following spread: This little bear will learn how to become a bear from the mother and then from the environment. The loss of the mother would wreak havoc with this apprenticeship.

The story takes place on Wrangel Island in the Arctic Ocean at a temperature of -22 °F. Kennan Ward is an expert on bears. In search of a bear's den, he found himself face to face with an orphaned bear only three months old. He decided not to interfere and to return to the base because of the blizzard. But this was without taking into account the little bear's determination to survive. The temperature had now dropped to -40 °F and in spite of the efforts of Ward and his teammate to dissuade the young animal, the bear cub followed the two men as if he were following his mother. "Young, inexperienced mothers often walk too fast for their cubs and lose them as a result. Sometimes too the last-born of a litter of three is deliberately abandoned because his mother thinks that she does not have enough milk... Moved by a survival reflex, the cub followed the first living creature he encountered. He stuck to me like a leech."

It is not unusual to see humans and even wild or domestic animals take on the role of adoptive parents. The mechanism of imprinting makes this bond possible and affects all living creatures at a precise moment in their life. During this period, known as the "sensitive" period, which is more or less long depending on the species, the animal's brain memorizes its environment so as to adapt itself to it. It is this phenomenon that makes it possible to tame and socialize animals. But if an animal is separated too early from the mother, it can cause mental problems in the young animal who then considers human beings as fellow creatures and may therefore try to seduce them or attack them!

Time for independence

Opposite: Only the dominant female in a pack of wolves is allowed to give birth, but then the whole social group looks after the young.

How do young animals decide to leave their mother? A lioness, a she-wolf, a tigress, a fox or a chimpanzee mother will often allow herself to be invaded and "tyrannized" by her young during the first few weeks, or sometimes the first few months or years of life. But the older the young become, the less their excesses and liberties are tolerated. So, at different ages depending on the species, mothers start to discipline their offspring by biting, hitting or threatening them. Shaken by this, they learn the limits imposed by their mother but without calling the maternal bond into question. When the moment of independence arrives, the young leave their mother and join the ranks in the social group; or, depending on the species, they move elsewhere, as is the case with two-year-old wolves, who leave their original pack and sometimes travel hundreds of miles before settling in a new place to start a family. For many animals, however, including elephants, killer whales, monkeys and crows, these maternal bonds endure.

Jane Goodall says that among the chimpanzees of Gombe in Africa, these bonds have always been strong and lasting. "Not only between mothers and their offspring but also between brothers and sisters [...] Brothers become good friends as they grow up and often stick together in the event of social conflict or physical aggression." Even among solitary animals, such as most of the big cats (cheetah, leopard or tiger), it often happens that when mothers and their offspring come across each other, they recognize and tolerate each other even after a long period of separation.

Affective submission

When the herring gull gives a cry of alarm as it flies off because there is danger, his offspring disappear among the rocks. The same happens with the northern lapwing, warned by his parents' alarm cry when a predator gets too close. In both cases the adults are trying to attract attention towards themselves. This behavior of birds trying to protect their offspring in this way was mentioned long ago by the philosopher Aristotle (384-322 B.C.). It was later described by the animal behaviorist Nikolaas ("Niko") Tinbergen (1907-88) who won the Nobel Prize for Medicine with Konrad Lorenz and Karl von Frisch in 1973.

It has also been observed that offspring brought up by and subjected to the authority of their dominant and protecting parents are more easily influenced by the society in which they are living, compared to young animals brought up with less care and therefore more apt to defend themselves on their own. Thus young wild boar will obey their mother and also the old wild sow leading the herd who makes all the decisions for the group and disciplines the recalcitrant young animals.

Parental solidarity

Opposite: Until the age of one or even longer, male and female wild boars rarely leave each other and often sleep huddled together.

There are numerous species where the whole family looks after the young. This is the case with gibbons and coyotes where brothers and sisters, uncles and aunts help their parents protect and bring up the new generation. Male and female belugas (white whales also known as "sea canaries") live separately within a large group. The females bring up their offspring together, breastfeeding them for about two years. Studying wild beavers in Quebec, François Patenaude observed that the one-year old animals helped their parents groom and feed the younger offspring with as much patience and care as is seen among the wolves within the pack. The solidarity within groups of female big cats and lionesses (whose reproduction cycles are synchronized) is well known. The babies will suckle from their mother or aunts without discrimination.

This cooperation also exists among female leaf monkeys who join forces to protect their young from the attacks of the males.

Pigs, whether domestic or wild, live in herds organized in a strict hierarchy and ruled by a matriarch. Until the age of eight months the young may be fed and protected indiscriminately by any of the other females in the group and when one mother is absent, another one takes over.

Young penguins are grouped in crèches to protect them from the cold, watched by one or a number of adults. Evolutionist biologists interpret this behavior as a genetic reaction aimed only at ensuring the survival of the species in case the parents should disappear. In contrast, many field ethologists speak of affective motivation and behavior, based on emotion.

Happiness exists in nature

Play, laughter, and enjoyment were thought of as the prerogative of the human species until it was discovered that animals also experience pleasure. As with human beings, play has a social function and playful activities often prefigure significant behavior patterns in adult life. Pursuing pleasure for its own sake also occurs in the animal kingdom, showing that animals are not prisoner of their instincts.

At play school

Opposite: Playing with this inedible sunfish takes a lot of energy out of this sea lion whose only aim is to experience pleasure.

Play occurs before birth. Like human babies who suck their thumb and move their hands in their mother's stomach, the little antelope kicks her mother and the dolphin sucks his fin and capers about as if training for later life. Play then continues long after birth, thus contributing to the development of social behavior and familiarization with the rules needed for living in society, and helping the individual become integrated in the clan. Animals who do not develop this interplay with their fellow creatures will ultimately be less good at the art of seduction and social relations when adult. A young creature who has not learnt the art of play will grow up a timid adult.

Playing helps the young animal to develop reflexes and endurance as well as muscles. For instance, the favorite game of young sea lions consists, as it does for their parents, in throwing and catching algae in the water, an effective way of becoming more familiar with an element that will play such an important part in their life. Animals who play are also the ones who will handle matters best when faced with a problem.

Good
players

Opposite: For these distant descendants of the albatross who decided to trade their wings for flippers, nothing is more fun than diving into the water one after the other!

Following spread: To satisfy a vital need to play, polar bears make the most of their environment and enjoy sliding on the icy snow.

Play is not the prerogative of humans, nor is it specifically reserved for young animals. It enables adults to express and maintain their bonds with their fellows. Social mammals demonstrate their mutual interest by friendly fondling that can be seen as play but also as more than that: they play for enjoyment and to laugh at the other. Adult sea lions have been seen playing with starfish or pulling iguanas by their tail in the water, thus preventing them from getting back to the jetty; crows enjoy sliding on the ice with their legs and stomach in the air, and dolphins sometimes play ball with sunfish and squids.

Dogs, badgers and polecats organize clever games of hide-and-seek where each side looks for the other and then reverses the roles. There is no difference between a cat playing with a mouse and the killer whale weighing several tons throwing penguins and seals into the air before eating them. An interesting point is that domestic rats burn up almost as many calories in enjoying themselves as human beings (about 20%)!

So many pleasures

Opposite: In Japan adult macaques often play with the snowballs made by their offspring.

When a female chimpanzee gave birth in a zoo in Holland, one of the working locations of the famous primate expert Frans de Waal, the whole troop became very emotional. A euphoric happiness spread to all the members when the nearest female screamed excitedly, then expressed her joy by hugging the two chimpanzees standing near her.

Less noisy but generating an equally infectious sense of well-being is the pleasure enjoyed by horses when being groomed, an operation that, according to expert Claudia Fey, has a calming effect on them and lowers their heart rate. More malicious was the provocative enjoyment of a young tiger, having gleefully uprooted the young trees just planted by his breeder, the American tamer Rick Glassey. Having committed his offense, the animal hid in a corner, covering his eyes with his paws! Each species has its own favorite game. Japanese macaques like to make snowballs and then sit on them.

For the pleasure of it...

Preceding spread: The friendly wrestling of these adolescent pandas (born in captivity) contributes to their development.

It was while observing chimpanzees looking at a sunset, listening to the lapping of the water or dancing in the rain, that the primate expert Jane Goodall became convinced that our cousins have a sense of beauty and in particular of enjoyment. They are not the only ones. It may be reasonable to doubt that a fly can experience a feeling of cheerfulness, but what about the dog, his leash in his mouth, leaping about at the thought of going for a walk with his master, or the obvious pleasure experienced by young pandas in their friendly wrestling matches? Undoubtedly they are all showing enthusiasm and pleasure. According to the ethologist Boris Cyrulnik, "We share with animals the same emotions of sexuality, hunger, protectiveness towards children, anger, despondency, suffering, and also joy and pleasure. Our molecules and hormones are identical, as is the way of handling information. The only difference: they are not triggered in the same way.

"If you are shown a picture by Picasso, his view gives you pleasure because he is your favorite painter. But you may also show it to your horse, and it's a safe bet that he will be indifferent. Unlike him, in this case your emotion is triggered by an artistic representation. On the other hand, offer him a field of grass instead of a painting and his pleasure will be as intense as yours, because it is the limbic brain which we have in common which supports this wonderful sensation of pleasure. Instead of art, he delights in the pleasant taste of grass!"

Educational recreation

Following spread: When playing, young polar foxes use a body language that they will use again to communicate with each other as adults, thus making such exchanges easier.

By play-fighting and pretending to submit, young foxes learn to control their aggressiveness, repeating rituals that in adulthood will prevent them from killing each other. The solitary games played by the little tiger cubs anticipate the adult hunting that they will take part in from the age of three. The jousting of young elephants, forehead against forehead, and young kangaroos boxing with each other, simulate the fights that males will later have to engage in to defend a territory or conquer a female. The play-fighting of the little wolves will determine the dominance of one over the other, that will enable them to establish a hierarchy and live in a pack.

The young whale prefers to swim and leap close to its mother, the kitten opts for a cork or a leaf to play with, while young antelopes and hares run races that will probably save their life in adulthood. Female cats bring mice to their young who by playing with them learn to improve their first clumsy hunting techniques, while foals-who do not have to learn to hunt to live-play at running away bucking. The games depend on the type of society to which the young belong. Young lions coordinate their efforts to organize an imaginary hunt, but young foxes, who will be solitary hunters, are less ready to share the things that interest them with their siblings.

Civilized
players

Domestication has made animals smaller, less reactive to danger, and in some cases more playful!

Thus, pet dogs have developed a juvenile behavior. Depending on their master for food, they behave like immature young wolves. They crawl, whine, play, bark-barking in wolves and coyotes disappears as they become adults-and beg for food. They can be jovial and provocative.

All the same, while domestication has made dogs more juvenile, their humanization has paved the way for a long-term apprenticeship, enabling them to develop further and perform intellectually far better than their fellow creatures in the wild, who still have to solve the immediate problems of survival, such as looking for food, finding shelter and so on. Yet, according to ethologist Elisabeth Marshall Thomas who has studied packs of dogs for over thirty years, they are very creative and particularly good at inventing games, such as group play, hide-and-seek, and so on. Cats who became close to humans approximately 6,000 years ago, still play with their prey or anything that looks like it with evident happiness. Everything is an excuse to remain alert and stimulate the senses.

Laughter is not the sole prerogative of humans

Opposite: The name "orangutan" comes from the Malay orang hutan, meaning "forest person." Here a large individual explores the forest like an acrobat.

Enjoyment is apparent in the expressions of the great apes, whose face muscles are highly developed. They smile, they burst out laughing, and they double-up with mirth. They make fun of one of their number by parodying them, provocatively, affectionately, or when they play hide-and-seek just by putting leaves over their eyes. In the words of Jane Goodall, who has studied them for 40 years, "Chimpanzees tickle each other or roll on top of each other while play-wrestling. These games are often accompanied by laughter, to such an extent that frequently adults cannot resist the temptation to join in… They know how to enjoy themselves just for pleasure. How many times have I seen a baby tickle the nose of an adult female […], the strange game where an older offspring holds a stick that he snatches away every time a younger one is about to snatch it? The little one cries, the older one laughs […] Like us, they play because it is fun."

All the while they make sure that they never show their canines, because this would be interpreted as a threat. The higher the chimpanzee's status in the group's hierarchy, the more discreet the smile, in contrast with the young whose joy is clearly visible on their faces and in their voices. When great apes laugh they produce a husky sound, very similar to the human laugh, but somewhat limited because the voice box is different. This direct comparison between humans and great apes is possible because of the facial expressions they have in common, but it does not exclude the possibility that other species may laugh as well. Indeed, some veterinary surgeons maintain that cats, dogs, and even rats are capable of laughing.

The pleasure
of lying

Preceding spread: A large ape on reprieve. Even his clowning will not move the people who are inexorably destroying his habitat, the Indonesian forest.

Some superior animals manipulate the emotions of others by mixing games with deception. Apes are thoroughly capable of lying. So a young baboon may have his eye on a tuber that a female is digging up. Suddenly he cries out in distress, thus attracting his mother's attention. To the young baboon's great satisfaction she chases away the female, and he then innocently picks up the tuber. Dolphins copy other animals, imitating the noise made by the bubbles escaping from divers' aqualungs. Crows also imitate the noises they hear. In studying birds who can talk, scientists have discovered that, depending on the species and individuals, these creatures are able to verbalize their emotions exactly like the great apes who have learnt sign language. In this way a rather bored parrot once tried to being a little excitement into his life by shouting out of the window overlooking the street: "Help, help, murderer!" The neighbors called the police who broke down the door of the apartment to the great pleasure of the lonely parrot.

Playing
in their dreams

Opposite: All animals with a developed nervous system of the brain can experience pleasure, but only mammals and birds are able to dream.

The young of invertebrates and cold-blooded vertebrates never play. Play is the prerogative of mammals and birds that made their appearance on earth 180 million years before man, warm-blooded creatures with stable body temperatures. The more immature the newborn, the greater the need to familiarize himself with his environment through play and through dreaming, activities that will contribute to his development. Young animals' dreams consist of representations of images that correspond to what they will learn for their later life as adults. Sleep consists of two different phases that alternate, REM (rapid eye movement) sleep and NREM (non-rapid eye movement) sleep. In REM sleep there is a cerebral alertness known as "paradoxical sleep," corresponding to the shallowest sleep, when dreaming occurs. The predator dreams a lot because his position enables him to sleep peacefully. He adopts a posture of lying in wait, expressing rage or fear, he appears to pursue imaginary preys or he sets off on reconnaissance expeditions. A cat has about 200 minutes of REM sleep a day in periods of six minutes and this continues throughout its life. The daily REM sleep of human beings is about 100 minutes in sequences of twenty minutes. Species with a high level of REM sleep are also the most playful. A chicken dreams for only six seconds and its chicks, fully developed at birth, are not very playful because they have absorbed much from their mother; since they do not have much to learn from their environment, this apprenticeship is enough for them.

Animal morality

Animals are capable of a wide range of behavior, including cooperation, mutual help, altruism, intention and sharing, that is very close to human morality. This has often been observed in the social groups of a species, and in certain situations with animals who are primarily solitary. From emotional infectiousness to empathy, the emotional brain plays an important part in this and shows that our human moral intentions are an animal legacy.

The legacy
of friendship

Preceding spread: Foxes have an innate sense of cooperation and mutual assistance. They create bonds when living in a group or as a couple. But it also sometimes happens that they are solitary.

Did friendship come into existence in the course of evolution, in the same way as other behavioral phenomena such as living in a society or parental care? Boris Cyrulnik explains it as follows. "In the world of genetically determined species, namely those species with a simple nervous system such as snails or fish, the level of bonding is relatively modest. There is a certain attraction towards the other that corresponds to chemical, tactile or visual signals of attraction. The creature responds to stimuli. Of course we cannot use the term 'friendship' in this case.

"On the other hand, in the case of a creature with a more complex nervous system-a bird or mammal-(with a prefrontal lobe associated with the memory circuits), commitment to a fellow creature outside of sexual attraction does occur. The creature responds to what it imagines and no longer to what it perceives. Here we see the introduction of a process that is a preparation for what we humans call friendship. The big apes are perfectly capable of empathy, arbitration, solidarity, consolation, reconciliation and friendship because they are able to picture the other one's world, which is something the sea slug is unable to do."

Contagious emotions

Following spread: In the wild, koalas are solitary creatures and extremely irascible. In captivity, the young deprived of their mother huddle together to ease their anguish.

Empathy means that an individual is receptive to the emotion conveyed by another individual and is concerned about it. This is quite a frequent phenomenon in social groups. At the Milwaukee Zoo in Wisconsin, a blind female bonobo is being looked after by a male who takes her by the hand to guide her along the corridors of the building. Being receptive to another can also take other forms. Maureen Frederickson, an expert in animal therapy, mentions the example of dogs who are able to anticipate the moment when a diabetic is about to fall into a coma. It appears that through smell dogs are able to detect the increase in acidity in a person's blood, the chemical process that precedes a diabetic coma.

Ethologists and veterinary surgeons have also observed that an owner's way of thinking can influence a dog's behavior. For instance, some people expect their dog to protect the house. They develop a fear associated with the surroundings that is perceived by the animal. Influenced by this emotion that is perceived through various channels, the dog then adopts a threatening attitude that his masters interpret as behavior defending the house.

Marthe Kiley-Worthington, an ethologist specializing in horses, explains that these animals are also aware of our emotional state. "A nervous person can make a horse nervous and suspicious, which in turn increases his feeling of fear. The horse receives and interprets a lot of information about our mood and intentions."

Compassion

For many years, Jane Goodall has been studying several groups of chimpanzees in Gombe in Africa. Our primate cousins have surprised her on many occasions with their human-like behavior, and they have also revealed their sense of compassion, as is illustrated in the story of Mel and Spindle.

Mel was three years old and had just lost his mother. He had no older brother or sister to take care of him and seemed condemned to die. But suddenly Spindle, a twelve-year-old adolescent who was not related to him, decided to take him under his wing. As time went by, the two young chimpanzees become inseparable friends. The older one waited for the little one when he fell behind, protecting him from other males, and letting him climb on his back or cling to his stomach as a mother would do with her young. This friendship lasted for over a year and it enabled Mel to survive. Such a phenomenon is also common among elephants. When a young elephant loses his mother, he is adopted by the rest of the herd.

Beyond
differences

Following spread: Gnus and zebras migrate together in the plains of Kenya and Tanzania in Africa. This reduces the risks and joining forces helps them fight off predators.

In his treatise on the cleverness of animals, written in the first century A.D., the historian and philosopher Plutarch (46-120) wrote: "The dolphin has no need of man and yet is man's friend and has often helped him." Since antiquity, stories have been told in which a dolphin supports a drowning man and carries him on his back to the shore, helps fishermen haul up their nets or guides ships through reefs. In the late 1980s, an Australian surfer attacked by a shark was saved by dolphins who protected him from the predator. The American psychoanalyst Jeffrey Masson tells the story of an old elephant who persisted in helping a rhinoceros stuck in the mud in spite of being attacked by the parents of the young animal. A female gorilla called Binti Jua, an inhabitant of Brookfield Zoo in Chicago, Illinois, helped a small boy who had fallen into the enclosure she shared with other gorillas. She went to him and protected him from the other gorillas before taking him to the keeper's door. In variations on the same theme, a bird fell to the ground in the chimpanzees' cage of the zoo in Basel. To the keeper's amazement, the sparrow was surrounded by the big apes and passed from hand to hand to the astonished keeper. In Chester Zoo it was a female bonobo who helped a stunned bird to fly off again. She climbed to the top of a tree and after gently unfolding its wings, she launched it into the air.

Social
assistance

Preceding spread: Gregarious by nature, lionesses help each other hunting, suckling their young, looking after them and keeping them at a safe distance from hyenas and lions from other packs.

Helping others is a common occurrence in the animal world. Female dolphins form coalitions to free other females kidnapped by groups of male bachelors. In all societies, assistance between its members is a common phenomenon, especially towards the young. Like the emperor penguins, flamingos organize collective nurseries while young elephants, lions, macaques, cape hunting dogs and jackals are supervised by family helpers. An explorer, Peter Freuchen, watched the determination with which a family of six wolves tried to free a wolf cub from a trap. When danger threatens, musk oxen form a defensive circle around their young to protect them. Avocets and wading birds have an unusual way of showing their altruism. When danger threatens, the male starts limping and fluttering a wing to attract the predator in the hope that it will spare his young. This tradition of helping each other also occurs in relation to individual adult animals. The Canadian biologist Paul Paquet observed the behavior of a pack of wolves where an old wolf too old to hunt was fed regularly by members of his group in exchange for looking after the young cubs in the group. Dolphins have been seen tearing at a fishing net to free one of their number who was caught in it, then guiding him to the surface to get his breath back. Sperm whales and other whales surround harpooned victims even at the risk of losing their own lives.

Surprising
affections

Opposite: Association between different species-as seen here between the dog and these leaf monkeys-shows that this type of behavior is not rigidly defined.

There are some remarkable stories. In Saint-Pierre de Bailleul in Eure, Normandy, a fallow doe lived for months with a herd of cows, fleeing as soon as it noticed human beings approaching but following the herd wherever it went.

In a nature reserve in Kenya, a lioness adopted several young oryx one after the other. This shows that there are elements of freedom in behaviors considered predetermined.

Plutarch tells this story: "A tiger who had been given a kid to eat starved for two days because he refused to touch it; on the third day, being famished, he expressed his desire to go back to his pasture so violently that he destroyed his cage: he did not want to eat the kid."

At the end of the 18th century the menagerie at Schönbrunn in Austria became the proud owner of a male Bengal tiger. French visitors told the following story, which was reported in the museum archives in 1806: "It is usually fed butcher's meat but when it is sick (it suffers from a type of ophthalmia), it is given live young animals, the warm blood of which helps it get better. A few weeks ago it was given a young butcher's dog [...] which, having recovered from its initial fright, went to the tiger and licked its eyes; the tiger felt so good about it that, forgetting its passion for live prey, not only spared the animal but showed its gratitude by stroking it [...] Since then, the two animals have been living together in perfect harmony and, before touching its food, the tiger always waits for its friend to eat the best pieces."

Long-term friendship

Preceding spread: A lion who leaves his pack to find another territory and start a new pack sometimes roams with a brother or fellow creature he meets on the way.

There are groups of bachelors in all social groups. Young lions sometimes wander for years before claiming a territory by driving away an old male from a harem of lionesses that they take over. They share their prey and help each when there are problems. In the world of baboons males leave the pack to found new families. When one of them arrives in a new pack, he strikes up a friendship with a young female. She is the one who will help him integrate into the pack. He seduces her by offering her gifts. Fruit, pieces of meat, a young antelope-the game requires determination and patience. Friendships that started in those conditions can endure for months or even years.

Solidarity

Following spread: When under attack the   musk oxen launch into battle.

In charge of a cell fighting against the trafficking of pets, Brigitte Picquetpellorce recalls an observation she made in Tahiti where abandoned dogs fill the streets: "One was a young dog, rather beautiful and looking like a little beige griffon. The other was an old mongrel, white with a black spot on his back. They crossed the street on a pedestrian crossing at a red light. The young dog went first, stopped in the middle of the street, let the old dog pass, and then followed him. When they begged, it was always in front of the same restaurant because they knew they would not be chased way. The young dog, clean and cute, begged for food and took it back to his friend who was waiting out of sight." Rémy Chauvin tells how coal tits responding eagerly to their hungry young when they are demanding food are very harsh to other fledglings who are not part of the nest. "And yet, remarkably, this harsh behavior is only superficial, because this young outsider who was chased away while he had parents could be adopted if he were lost or orphaned!"

To protect themselves against wolves, musk oxen regroup into a defensive circle or square with the youngest animals in the middle. The astrophysicist Carl Sagan described a laboratory experiment that consisted in forcing monkeys to chose between giving electric shocks to their fellow monkeys or being deprived of food. Almost all the monkeys refused to hurt their fellow creatures to the point of not eating for two weeks! Experts in evolution interpret this altruistic behavior as an instinctive manifestation of survival. In other words, according the geneticist Richard Dawkins, friendship among animals can be explained by their desire to preserve, from a genetic point of view, individuals of their own species. How then can we explain the help given by animals to individuals of other species?

An eye for an eye, a tooth for a tooth

Aggressiveness and anger are basic emotions as is hunger, the driving force behind hunting, and they play an integral role of the animal world. On the whole, animals resort to rituals to avoid bloody fights, but it also happens sometimes that members of the same species wage war among themselves.

Brutish behavior

Preceding spread: Relationships between horses run deep and, as in any society, there are as many affinities as there are antipathies.

Aggressiveness and sexuality are often linked. Konrad Lorenz used to say that rape did not exist in nature. Yet the sea elephant sees his partners, five to ten times smaller than him, as sexual objects and the violence of copulation if the females resist looks very much like rape. If they try and flee, they are caught and bitten. When the males are unable to satisfy their sexual desire, they turn to the young of both sexes with extraordinary violence. Copulation can be brutal in many species, including orangutans.

The sexual behavior of large dolphins also lacks courtesy. It is not uncommon to see two groups of unattached males join forces to kidnap the fertile females of a third rival group. Having committed the offense, the female is then set upon by two or three individuals who do not hesitate to bite and hit her to make her surrender to their sexual demands. In studying the deep bond between mare and stallion, the ethologist Marthe Kiley-Worthington noted that the serving of a mare in a stall, covered without preliminary by a stallion (the usual method in livestock farming) traumatizes the female. "They tremble, sweat profusely, and become anxious and recalcitrant when they are forced to go back to the stall... This practice is well and truly close to rape."

Rivalry in the land of wolves

Following spread: Wolves are well known for patrolling the boundaries of their territory as do the big apes. Any unknown individual will be driven away.

In 1995 fourteen wild wolves, captured in the Canadian Rockies, were released to the north-east of Yellowstone National Park by scientists of the Gray Wolf Recovery Project. They were then joined by sixteen wolves captured in British Columbia. Soon they grouped themselves into three packs that were named Rose Creek, Druid Creek, and Crystal Creek. The Druid Creek pack was very aggressive, chasing away any intruder venturing on its territory.

At the beginning of 1996 the Druid Creek gang declared war on the Crystal Creek gang. They killed the dominant male and injured his companion. A few weeks later, the Druids attacked the other gang, the Rose Creek pack. Then the dominant male of the Druids was shot and killed by a hunter at the park's boundary.

At the same time, a young male belonging to the Rose Creek gang left his original pack to found his own. The young male was determined. One day he passed the female of the Druid pack and walked up to her with amazing self-confidence. At first aggressive, she threatened him and then tested him by jostling him while the rest of the pack surrounded him. This went on for six hours! But the young male resisted, did not become impatient, nor did he show anger or fear. Then suddenly everything changed: the female was conquered, becoming sweet and affectionate. This was a signal for the rest of the pack who rushed towards him not to kill him but to recognize him as their leader and to show him their allegiance by licking his chops as wolf-cubs do. When two years later another young male of the Rose Creek pack also decided to cross the territory of the Druids, none of them attacked him.

The
scapegoat

Previous double spread: Because she behaves differently from the others, a lioness may be rejected from the pack she belongs to and forced to lead a solitary life.

In the case of conflict between two groups, the defeat of the losing pack often leads to changes in the social structure of the group. The males establish a new hierarchy and choose a scapegoat. This will turn them into dominant individuals for a time and thus appease their frustrations. Life can go on. Cows coexist because their life is based on strict rules that enable them to solve most conflicts through particular postures and the use of vocal and olfactory signals. Nevertheless, if 250 cattle are put in a closed pen, the dominant males will relentlessly attack the weaker individuals: such a large number makes it impossible to build a social hierarchy. The solution is simple. By reducing the number to fewer than 200 head, the problem will disappear. Animals that live in groups-including cats, lions, chickens, wolves, zebras and monkeys-very often choose a scapegoat on which they can vent their aggressiveness. This helps them balance tensions, bring harmony to the group and avoid collective fights. The victim lives in a state of permanent fear, eating very little or not at all, trembling and existing on the fringes of the group. If the individual dies or if he is removed from the group, another animal is usually selected as the new scapegoat.

Fascinating
executioner

Opposite: Predators always kill without aggressiveness. But in the case of a conflict between two or several individuals of the same species there will always be aggressiveness.

A monkey went up to an antelope, took her in his arms and played with her. The little antelope did not struggle but seemed fascinated. Then suddenly the monkey killed her and crushed her skull to eat the brains. The same happened with a leopard who caught a baboon and then pinned him down without using any force. The monkey was completely relaxed and appeared to be offering his throat without attempting to flee. These two observations made by ethologist Boris Cyrulnik are not unusual. In fact, they are quite common and corroborate the theory put forward by some experts, that the fascinating charm of the predator is believed to prevent any pain being felt. In other words, there is an "adaptive" mechanism that enables the animal to die without suffering. This phenomenon is also seen among humans where victims of a major trauma literally become anesthetized, undergoing a traumatic and emotional stupor that we share with animals. If the predator stops his attack, the prey's stupor fades away, the victim regains awareness and walks off, now insensitive to the bewitching charm of the executioner.

Signals
of distress

Preceding spread: Charged by an elephant, this lion is indubitably expressing his fear. The fact that he is lying down indicates, if not submission, at least that his courage is being put to the test.

Bodies express the emotions of individuals. To coexist in harmony, wolves must submit to the dominant individuals by adopting a behavioral ritual that controls and inhibits the aggression of their hierarchic leaders. They flatten their ears, lower their tail, avoid their gaze, and approach the dominant individual like puppies, whimpering to lick his chops. The highest ranking animal in the hierarchy understands and accepts the importance of his role, which consists in uniting the group and arbitrating in conflicts. The same happens with lions, who offer their jugular vein, or seagulls and crows who turn away their beak. A young horse getting a little too close to a mare belonging to a stallion will quickly adopt a whole series of conciliatory postures and displays of body language using his mouth, neck, ears, and tail to soothe the dominant horse's anger. The most important advantage of animal hierarchy is precisely the neutralization of aggression. For instance, male sea lions calm down conflicts among the females with a few ceremonious greetings. A monkey curls his lips and advances towards his aggressor with the palms of his hands facing upward to reduce the other's aggression and stop him in his tracks. In ways like these, most animal conflicts are resolved in a peaceful manner.

Animal violence

Following spread: Fights between hippopotamuses (named from the Greek meaning "river horse") are impressive for their violence. Sometimes the outcome is fatal.

When food is scarce, animals do not hesitate to sacrifice parents and siblings. For instance, female prairie dogs bury other gestating females alive in their burrows. This is a drastic but effective way of limiting the population. In the case of some birds whose eggs hatch at intervals of several days, the lastborn chicks are used to feed the older ones. By killing her neighbor's young, the crow stops her from encroaching further on her territory.

Yet animals are by nature neither cruel nor violent, and this amazing aggressiveness should not by judged by human standards of morality. In a system of competition and social order, evolution has selected this behavior in a species as a means of enabling its members to make use of the resources of the environment, to ensure reproduction, and to maintain equilibrium within the group. So in a group of hippopotamuses, a species closely related to the whale, every male must submit to the dominant male at the risk of losing his life.

Other factors may intervene. The robin is stimulated more by color than by movement. A few red feathers abandoned on his territory will trigger a furious reaction and angry trilling to express his displeasure, sounds interpreted by humans as melodious song.

An aggressive environment can also lead to biochemical changes. Thus, a study of thirty-five generations of mice has shown that when individuals are put in an environment that favors aggressiveness, violence becomes hereditary. Three or four generations are usually enough to produce combat mice!

Bonus for You!

Your public garden membership includes a one-year subscription to *Better Homes and Gardens*®!

Better Homes and Gardens®—one beautiful real-life garden. What to plant now, where to plant it with easy projects to give your garden a boost.

BROUGHT TO YOU BY

A P G A

INCLUDED WITH YOUR PAID MEMBERSHIP!

If you are a current subscriber and would like to extend your subscription, check here ☐

APGA Member
Garden Name: _____

Your Name: _____

Your Address: _____

City/State/Zip: _____

Email Address: _____
Yes, email me with news and special offers from your family of publications.

BHG NAPGA03

FILL OUT & MAIL THIS CARD NOW
TO CLAIM YOUR SUBSCRIPTION!

BUSINESS REPLY MAIL

FIRST-CLASS MAIL   PERMIT NO. 120   BOONE, IA

POSTAGE WILL BE PAID BY ADDRESSEE

Better Homes and Gardens.

PO BOX 37815
BOONE IA 50037-2815

Fighting rituals

Opposite: Male kangaroos often fight over a female, grappling and hitting each other just like two boxers.

The history of animal fights organized by humans (dogs, tigers, cockerels, bulls, horses, camels… and even crickets!) is full of examples of animals refusing to fight. The last tiger fight, which took place in Marseille in 1908, saw two wild animals trying to get away from the bull they were supposed to fight. A hundred years earlier in Vincennes, a cow got the better of a tigress who was refusing to fight. Today, in the Guangxi region in China, when a stallion fight is organized, a mare is often used as a stake. But even in these situations animals use a wide range of ritual acts to avoid death and serious injury.

But rats are the exception: aside from a certain number of individuals, every mouse or rat is marked by the scent of the group. If one of the individuals is absent from the group for a few days and as a result has lost the scent, on his return he will be perceived as an outsider and attacked, or even killed, by his fellow creatures. Anger, a fundamental emotion, is the response to a threatening situation. When frightened or angry, an individual secretes adrenaline that accelerates the heartbeat, stimulates breathing, and prepares the body to flee or to attack.

Offensive and counter-offensive

Preceding spread: The hunter hunted... This cheetah is fleeing from an angry warthog that is refusing to become a prey.

With animals as with humans, the main purpose of war is to defend a territory. In the case of the big dolphins, males form alliances. A group may ask a second group for help to attack a third one. Among mongoose, wolves, and rats, the groups face each other and advance towards each other like two armies. The smaller the territory, the bigger the aggressiveness within the same species. In 1974, Jane Goodall witnessed a punitive expedition by a group of male chimpanzees against a group established to the south of their territory. First they surprised a male who was held to the ground and beaten up with stones and kicked, then they proceeded to kill all the males and kidnapped the females whom they took home. Every day a silent patrol went to the territory of the enemy clan. The left-over food was checked and the nests were inspected.

There is another form of aggression: the counter-offensive of prey against the enemy predator. Danger is so strongly imprinted in many species that the individuals react with systematic harassment. So we see Canada geese pursuing foxes, jackdaws and magpies harassing cats, oxen forming an aggressive circle around a dog or a wolf crossing their pasture, saber-horned antelopes killing young lions, buffaloes trampling young leopards, and even young foxes standing up to bears!

Predators

Opposite: Male baboons often capture and kill young antelopes that they then offer to female baboons as a gift.

Emotion is contagious. When the anguish and stress emanating from wild or domestic herbivores arouses the aggression of wolves or dogs, the latter can kill not because they are driven by cruelty but because they are conditioned to do so. The imperatives that determine the behavior of a hunter are different from those of an animal defending its territory. To illustrate this, the Austrian ethologist Konrad Lorenz explained that the "buffalo killed by a lion did nothing to provoke the lion's aggression. A dog jumping on a hare has the same happy, attentive expression as that with which he greets his master or waits for some pleasant event." Completely different motives are in force when a female stork returns after migration to find another female next to her faithful companion and... kills her. According to ethologists, the killing of a prey cannot be seen as an act of aggression because it is motivated by hunger. Hunting is so deeply rooted in predatory species that they capable of doing it again if even they do not need to.

An experiment has been carried out in which eagles were fed and all potential prey was removed from their environment. After a while the birds became nervous and attacked anything that moved, such as rolling pebbles, and waving branches and leaves.

Pain and suffering

Animals have the ability to suffer. Although this has long been called into question, the evidence put forward by science today could very well affect the debate in the future: do we have the right to cause suffering to a creature that can feel emotions?

This is the important question because if animals have emotions, intelligence and awareness… what right do we have to exploit, destroy and torture them?

Sensitive animals

Opposite: Cows are sensitive beings who suffer all the time from the conditions prevalent in industrial farming.

While many studies have been carried out on both domestic and wild animals, researchers have on the whole shown little interest in farmed animals such as cows. John Watts, a scientist at the University of Saskatchewan in Canada, has turned his attention to this. He has demonstrated the seriousness of behavioral problems in these animals when under excessive stress. Dan Weary, a professor at the University of British Columbia, has highlighted the trauma and distress of young heifers taken to the milking parlor the day after giving birth to their calf, while the calf itself is left alone in a narrow stall. Studies carried out in Canada and Australia have shown that the presence of a person the cows are afraid of during milking can reduce their milk yield by five to ten percent. Stress induced by fear leads to a retention of the milk in the udder, caused by the disruption of the secretion of lactation hormones. These researchers are emphatic: "Almost everything in the way we treat cattle is abnormal. From a psychological point of view, they are wild animals crammed into artificial prisons. People are concerned about the treatment of animals in zoos, but they should also worry about the treatment of farmed animals because they are the same."

Emotional anguish

Preceding spread: The rapidity with which cats kill their prey helps to shorten their suffering.

Monkeys and elephants, who seem to have a primitive notion of death and absence, are very upset by the death of one of their number. But do farmed animals taken to the abattoir have an idea of death? Researchers are beginning to work on this subject.

It seems that animals are aware of the dead body of another and of the anguish generated by fear in particular, but not of death at the time. The cries of animals going to be slaughtered and the smells prevalent in such an environment add to their stress. The urine of a frightened animal contains components that induce fear in others. The suffering of one animal spreads to others. Through contagion, the emotional anguish experienced by their fellow creatures leads to increased fear in all the rest. This results in panic movements, refusal to move forward, attempts to flee, and so on. Pigs, for example, are very sensitive to their environment and form significant social bonds. They are therefore deeply affected by the breaking of these bonds.

Suffering
in silence

Following spread: Pigs are hypersensitive creatures and very fearful. In the last seconds of their life their cries can reach a sound level of 100 decibels.

The cerebral structures involved in suffering are part of the limbic system, in other words the emotional brain that exists in all mammals. Animals usually respond to danger by physiological, biochemical and behavioral changes, such as sweating, increased heartbeat, hairs standing on end, micturition and defecation. While extreme suffering manifests itself in the form of agitation, fear and cries similar to those arising in combat and escape situations, by contrast chronic suffering will result in silent withdrawal and deterioration of health.

It is true that suffering varies according to species, but recently scientists have discovered that fish also suffer. Robert Dantzer, veterinary surgeon at the French national institute of agronomic research INRA. has said: "We now know that animals share certain elements of awareness with humans, namely a mental representation of what they have to do and what they intend to do.

"If this embryonic awareness determines the ability to suffer, the animal has all the elements to experience suffering. An animal suffers when it is unable to express the repertory of its natural behavior. This definition leads us to reassess the entire system of industrial farming that forces animals to live in conditions that are not adapted to their biology."

While pain is a physiological reaction, suffering requires perception. Animals are able to experience pain and suffering because they have sensory perception and a memory. "I was taught that animals, like children, do not suffer," Boris Cyrulnik recalls. "When I was in my first year of studying medicine, we had to dissect live animals and when these cried out and put up a struggle, the professors assured us that they were not suffering, claiming that a squeaking bicycle did not suffer either!"

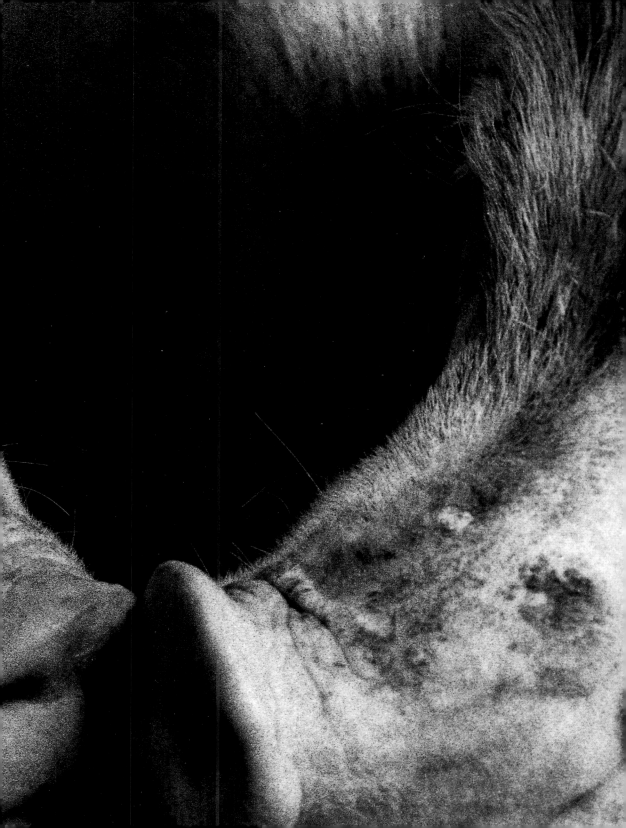

Sensory isolation

Opposite: Through boredom, an animal living in captivity may gradually lose all interest in life.

Since the early 1950s, a multitude of psychological studies have been carried out on animals. To this end, ever more cruel and disgraceful methods have been used to cause dementia, submission, a state of shock or depression in the subjects. Numerous experiments have been carried out in which animals are kept in complete isolation in cages or steel containers without hearing or seeing any sign of outside life for months or even years, in order to evaluate their behavior. These experiments have shown that after a certain period, the animal huddles up. Sometimes it will leap up and stand up wide awake, panicking and whining, as if aware of an unfamiliar presence, then return to its prostrate position. Experts believe that the reactions of animals in these experiments is comparable to hallucinations in humans when, as a result of sensory deprivation, the brain produces images as a mental protection.

A dog's life

Opposite: A dog often depends on his master for the best and often too for the worst.

According to many veterinary surgeons, domestic animals have become replacements for our affective deficiencies and our whims. The consequence is mental suffering, neuroses, and psycho-somatic illnesses that do not occur among their wild fellow creatures living in their natural milieu.

Dogs are particularly sensitive to the behavior of their master. Ethologists now describe them as affective sponges whose development is shaped by their master's state of mind. The animal adapts to its master's mood, sometimes happy, sometimes sad, and sometimes angry or indifferent. The case of the dog that has died and is then replaced by an identical one is a perfect illustration of this. The animal suffers profoundly from the comparison its master makes between it and its predecessor. This relationship, based on contradictory emotions, disappointment, and affection, causes behavioral disorders and physical problems: stomach ulcers, shaking, skin problems, licking wounds (the animal licks itself until it bleeds), sphincter problems, and so on. The dog will only get better if the master mourns his first companion before welcoming the second dog as an individual in its own right, and above all as completely different!

Gut fear

Alex was a grey parrot from the Gabon who was taught to speak by the American ethologist Irene Pepperberg. One day she took him to the veterinary surgeon for an operation. As she was leaving the anxious bird suddenly shouted: "Come here, I love you, I am sorry, I want to come home." Some chimpanzees, terrified at the sight of one of their fellow creatures paralyzed by poliomyelitis, fell into each other's arms and kissed each other to reassure themselves. Fear, an essential emotion in the face of danger, is perceived and expressed in different ways, depending on the species. It can affect all living beings, even a simple sea slug!

For instance, the particular size of its own space determined by each individual is a bubble that surrounds the animal as a function of its emotional state. The violation of this space usually trigger a reflex of fear that varies according to the situation. There are exceptions: like many other small animals, the young herring gull does not perceive danger itself; the awareness of danger is linked directly to the behavioral reactions of its parents. Among elephants, fear can be transmitted through infrasounds (low frequency sounds). For instance, an account was given of a group of African elephants that was warned of the massacre of another herd dozens of miles away. Then there is the example of rats so traumatized by the aggressiveness of their fellow creatures that they died of fear. On this subject Konrad Lorenz wrote: "It is unusual that one can read so clearly in an animal's expression such despair and fearful panic, combined with the certainty of a horrible, ineluctable death, as in the case of a rat that is condemned to be executed by other rats. It does not even bother to defend itself any longer."

Awareness of death

Following spread: Young elephants stroking the corpse of one of their number with their trunks.

The pain caused by the death of a member of a social group is a universal emotion probably shared by all living beings endowed with awareness. Elephants have a relationship with death that is described as strange because their behavior is so sophisticated. A female elephant whose baby has died will carry him on her tusks for days on end while another mother with her trunk will try madly to make her dying baby stand, cuddling him as she tries to feed him. The herd will gather round the dead body and, remarkably, will cover it with twigs, branches, grass or earth, a behavior that has been observed several times.

Cynthia Moss, founder and head of an elephant research center, recalls an old female elephant who every day crossed the center where skulls were collected for analysis, always stopping in front of the same one. "She would always touch it with her trunk, sniffing at it, touching it gently with her foot as elephants do when they are sad and pensive, as if she recognized it among all the others. It was the skull of one of her daughters, who had died two or three years earlier." She also describes the reaction of several individuals in a herd of female elephants (the males live in separate groups) when one of them died: "They acted as if they were mad; they kneeled down and tried to lift her, sliding their tusks under her back and head."

The lament
of the bear

Following spread: Polar bear cubs (there are one to four cubs in each litter) stay with their mother for two years. The death rate is high in the first few months of life.

Canada. Mid-November on the shores of the icy Arctic Ocean, just above Churchill. Daniel Cox, a professional photographer, was watching a female bear with two cubs, one of which seemed to be having difficulties in following her, although the mother continually looked back and called him. "She kept going back to get him," said Cox, "sitting next to him while he was recovering. Then she would set off once more, returning again after twenty or thirty yards…" The next day the photographer noticed that the condition of the little cub had worsened. The following day the cub no longer moved. The mother sat next to him and did not leave his side. A few hours later the little cub was dead. Several times the mother tried to lift her baby by grabbing him by the neck with her powerful jaws, brushing the powdery snow away and driving off another bear attracted by the smell of the corpse. She spent two whole days beside her baby before setting off with her surviving cub.

"How does a mother bear know there is no more hope?" Daniel Cox wondered. "This vision moved me deeply. I could feel the mother's anguish; her distress and probably her fear that the remaining cub might become the victim of the fangs of a hungry male."

Other similar cases have been reported, such as a young wapiti watching over the dead body of another wapiti killed by a coyote, and a female sperm whale carrying her dead baby in her jaws for several days. There are enough examples to contradict the statements of some philosophers who deny that an animal suffers when losing a mate or its young: "In animals, indifference to external suffering is complete," Jeanine Chanteur writes. "Behavior involving rescue is not deliberate […] If we observe the suffering of a human mother at the bedside of a terminally ill child and that of an animal mother, there is no possible comparison […] The animal mother does not worry about her young or even rejects it."

"If animals did not exist, would we find it even harder to understand ourselves?"

Buffon

It was in the oceans that the first living creatures appeared in the form of cells and microscopic algae. Then insects developed in the air before animals finally succeeded in conquering the earth and evolving into a fantastic profusion of species.

It was about three million years ago that Homo habilis, the first "skilful" human being, appeared. Sharing a common ancestor with the big apes, Homo habilis had a complex brain that was the result of an evolutionary process over 600 million years, consequently sharing with mammals and birds a vast biological and behavioral inheritance including a wide range of emotions that are vital in life.

We are not unique beings but we have a unique ability that animals do not have: that of asking ourselves questions about the future of the earth and the impact our behavior has on it. About a dozen species of human beings and hominids have existed, all but one of which have become extinct. We, Homo sapiens, are the last such species on a planet that we are relentlessly destroying: a quarter of all animal and plant species could disappear in the next fifty years. We must think carefully about this reality and feel humble in the face of the diversity of life, animal beauty, our common origins and the emotions we share.

Thanks

To Boris Cyrulnik for his generosity and patience under all circumstances.
I also warmly thank the researchers, artists, self-taught individuals, and most of all these photographers for their remarkable talents and knowledge of the animal world.
A big thank-you to Joan Le Boru, a longstanding supporter, and to Virginie Freyder for her unfailing enthusiasm.

Photographic credits

© David Bailey : p. 131.

© Bios/Theo Allofs : p. 139 ; Fred Bruemmer : pp. 116-117 ; Mathieu Laboureur : pp. 136-137 ; Cyril Ruoso : pp. 54-55, 111.

© Michel-Denis Huot : pp. 140-141, 143.

© Jacana/Anup Shah : pp. 45, 88-89.

© J.L. Klein & M.L. Hubert : p. 85.

© Michèle Le Braz : pp. 154-155, 158.

© Liberto Macarro : pp. 35, 65, 148.

© Minden Pictures/Joël Halioua Editorial Agency/Jim Brandenburg : p. 61 ; Mathias Breiter : pp. 58-59 ; Gerry Ellis : pp. 164-165 ; Gerry Ellis/Globio : pp. 78-79 ; Katherine Feng/Globio : p. 49 ; Michael & Patricia Fogden : p. 112-113 ; Heidi Hans-Jurgen Koch : p. 42 ; Michio Hoshino : p. 17, 18-19, 82-83 ; Mitsuaki Iwago : pp. 26-27, 46-47, 100-101, 102-103, 106-107, 132-133 ; Frans Lanting : pp. 22-23, 32-33, 50, 52-53, 73, 87, 91, 157 ; Konrad Wothe : p. 76 ; Shin Yoshino : pp. 36-37 ; ZSSD : pp. 62, 74-75.

© Florian Möllers : pp. 160-161.

© Vincent Munier : pp. 21, 96-97, 108-109.

© Natural Exposure/Daniel Cox : pp. 168-169.

© L.M. Préau : pp. 28-29.

© SeaPics.com / Joël Halioua Editorial Agency / David Hall : p. 70 ; Doug Perrine : pp. 14-15.

© Stock Images/Steve Bloom : pp. 128-129.

© Sunset/Horizon Vision : pp. 122-123.

© Art Wolfe/Joël Halioua Editorial Agency : pp. 126-127, 150-151.

Publishing manager JOAN LE BORU
Picture research VIRGINIE FREYDER, JOAN LE BORU, RÉMY MARION, KARINE LOU MATIGNON
Copy editor VIRGINIE FREYDER
Art direction SABINE HOUPLAIN
Graphic design SANDRA CHAMARET

Reproduction: QUADRILASER